# The First Raindrop

By
Lyna C Nath

*For Emma, my Little Raindrop*



Library of Congress Control Number: 2018902898

On a warm mid morning in autumn,
the sun's rays shone brightly over
the calm ocean water.

The sun heated the ocean water.
Some water droplets dried up
and floated high into the sky.

Over many days and nights, billions of water droplets joined together to form a huge cloud.

The dark cloud grew heavier and heavier as each day past until it could no longer contain itself.

A single tear fell from the angry sky that cried with thunderous roars.

Little raindrop looked down toward the ground and saw a fun place to land.

The first raindrop plopped,
dropped upon a golden leaf
of a sycamore tree.

It fell on a child's freckled face and trickled down the cheek and chin.

Little raindrop slipped,
dripped on the petal
of a thirsty
dandelion.

The first raindrop splashed on the concrete and made a mad dash between the sidewalk crack.

Little raindrop crawled, sprawled into the dark stormdrain.

It mixed, mingled with other droplets while it struggled past week old newspaper.

The first raindrop tangled with strange objects before freedom ahead at pipe's end.

It twirled in a whirpool
dance and swirled down
rusted, crusty city pipes.

The first raindrop poured over a powerful waterfall into a slower moving stream.

Little raindrop rushed over smooth river rocks and moved quickly with the flow.

The first raindrop
floated easily by
a busy beaver.

Little raindrop splished, splashed in the water with a couple of frisky frogs.

It swam in the currents
with a group of friendly
fish.

Little raindrop drifted lazily
by a turtle who was basking
in the early evening sun.

The first raindrop slid by reeds
and river birds as it happily grew.

It jumped and twisted joyfully while playing with its friends in the waves.

Little raindrop joined its new family of a billion, trillion other drops in a sea of water.

Made in the USA
Monee, IL
04 November 2026

67374645R00017